I0467928

Solar PV Powered UV Water Treatment:

How to Solar Power UV Water Sterilizing Systems for Drinking Water Onsite

by Christopher Kinkaid

Solardyne.com

Published by Solardyne, LLC
Portland, Oregon

ISBN-13: 978-1500472610
ISBN-10: 1500472611

Table of Contents

Preface

Sterilizing water is a big job. Solar electric (PV) powered UV water sterilizers, are an effective way to sterilize your water from local polluted sources, even brackish water, with safety, reliability, and no fuel-costs. Water found in nature is full of pathogens, which can cause disease and illness. Ultra-violet (UV) sterilizers kill 99.99% of all dangerous pathogens and renders your water potable and safe to drink.

The need for water treatment usually occurs far away from an electrical outlet. Remote sites and locations, as well as occasions of Natural or Man-Made disaster, often need water treatment onsite, but lack the equipment and power supply to power the water sterilizing equipment on location. Solar PV powered UV water sterilizer systems offer complete solutions for remote-site water treatment and sterilization. This Book focuses on UV water treatment from 4 Gallons per Minute to 43,000 Gallons per Day - all Solar Powered. Included are specific Solar Power Supply examples, with included Parts Lists, to power these UV water sterilizing systems at your remote and off-grid location.

Note: The Solar powered UV systems listed are for well, or surface sources of water which are brackish, and/or polluted. In the case of Salt Water sources of water, then Desalination equipment is required Before the UV treatment phase.

About the Book

This Book is written as a step-by-step guide to defining your solar water treatment project's "vital statistics," and choosing the right equipment to get the job done. If you have a specific solar powered UV Water Sterilizing project in mind, then visit the Solar PV Powered System Examples List located at the **Quick Guide** in **Chapter Eight**.

Note: The Solar powered UV systems listed are for well, or surface sources of water which are brackish, and/or polluted. In the case of Salt Water sources of water, then Desalination equipment is required Before the UV treatment phase.

The **Quick Guide** contains clickable links which take you to a specific UV Water Sterilizing System Total Daily Water Production, and the Solar PV power supply needed for operation.

The UV Water systems are defined by Flow-rates, and delivered Gallons per Day. Solar PV power supply examples are defined by the Gallons Per Day of potable water delivered. If you're sourcing your water from a Salt Water source, then you'll need a Reverse Osmosis (RO) System before the UV water sterilizer see Chapter 8. Chapters 4-7 deal with "Fresh" water sources like ponds, creeks, lakes and streams (brackish, or polluted), with Chapter 8 focusing on Salt water sources.

The UV Treatment systems listed in the examples are based on different Flow-rates. There are Four UV sterilizer systems including 4, 8, 12, 30 Gallons Per Minute. Each system will have several Solar Power Systems defined which power the UV system for 4, 8, 12, and 24 Hours per Day, respectively. Choose your Solar powered UV water treatment system based on your desired Flow rate, and how many Gallons Per Day you need to sterilize to best match your water treatment project. Included examples, range from 240 GPD to 43, 200 Gallons Per Day - all **without chemicals**, or **fuel-costs**.

Chapter 2, outlines the Step-By-Step process to define your UV Water Treatment system for your own system design, or to speak with an outside vendor. Use this process to determine the "vital statistics" of your system, and Size your UV System and Solar PV power supply easily.

Chapter 3 discusses the use of Solar Power Supplies, and how the listed examples are configured in this eBook.

Chapters 4 through 7 describe UV Water Treatment Systems and the corresponding solar PV power supply to deliver a specified volume of potable water, ready and safe to drink. System examples, include Solar PV Power supply parts lists describing the specific solar PV panels, and electrical components you'll use to operate your UV Sterilizer for highest productivity.

Chapter 8 discusses UV Water Treatment Systems for Salt Water Sources, with Solar Power supplies. Solar PV Systems are defined by total power and energy they can deliver for your load. In all cases solar PV panels will charge a battery bank to provide power and energy for the UV sterilizer any time day, or night.

This book "Solar PV Powered UV Water Treatment" was written to be a resource for planning and implementing a Solar electric (PV) powered UV Water Sterilizing system to deliver potable, clean, and safe water at remote sites. Ideal for remote cabins, homes, off-grid living, residential, commercial, Disaster Relief, or any location where there is no, or limited, local electricity, and the need for clean water is acute.

Solar power panels are an excellent power supply choice and enable water treatment systems to operate where no electricity is present, or to provide back-up should a local grid go down due to disaster.

About the Author

Christopher Kinkaid

Christopher (Toby) Kinkaid, originally from Portland, Oregon is the founder of **Solardyne.com**, **SolarQuote.com**, and **AlgaeToday.com**, and has worked in clean energy technology for over three decades.

Kinkaid, is the inventor of the "**Helyx**" Vertical Axis Wind Generator, the "**Mariposa**" Non-imaging solar concentrator PV module (continuous operation at Sandia National Laboratory since 1994), the **Solar Demultiplexer** optical solar concentrating lens (Dr. James/Sandia National Laboratory 1991), and the inventor of the original "**Solar Power Pack**" (Mother Earth News, "**Littlest Utility**" June/July, 2001).

Kinkaid, has been an official lecturer and presenter on clean energy technology around the world including "APEC", Bangkok, Thailand, 2003, "Energy Solutions World", Tokyo, Japan, 2003, The International Biomass Conference (IBC), 2010,

Minneapolis, MN, and the Algal Biomass Organization (ABO) Conference, 2010, Phoenix, AZ.

Christopher (Toby) Kinkaid, has appeared in interviews on KOIN TV, KGW TV, and "Sustainable Today" produced in Oregon, and has served on the board of directors for the National Hydrogen Association, in Washington D.C., 1993, the Japan Satellite Communications Company (JCNET), Fukuoka, Japan, 1994-95, and Algaedyne Corporation, Preston, MN, 2010-2013.

Kinkaid, presently serves as CEO of Solardyne, LLC in Portland, Oregon, where he continues his work in Solar, Wind, and Biomass Technology, applications, research, and development.

Introduction

The need for cleaning water is fundamental to life. Without clean water to drink, there is no civilization. Natural sunlight contains Ultra-Violet (UV) rays, capable of destroying dangerous pathogens found in water, by disrupting their cell DNA. Today, modern technology takes a cue from nature and uses high-efficiency UV light bulbs to irradiate polluted water killing 99.99% of all dangerous pathogens.

Irradiating your water with strong UV levels destroys these dangerous organisms, allowing you to source your water from local Wells, or shallow sources like creeks, ponds, rivers, and streams as a source of potable water.

Today, modern solar electric panels (PV) can power UV water sterilizers making clean energy available at your remote site, which are easy to install, cost-effective, and offers outstanding performance and reliability where it counts: day-by-day in operation. Solar PV panels are solid-state, have no moving parts, hermetically sealed from the environment, rated for extreme locations and often carry 25 year warranties making for a reliable power supply.

With proper design, and hardware choices, (the point of this eBook), solar powered UV Water Treatment systems are surprisingly productive purifying water from four Gallons per Minute, to tens of thousands of Gallons per Day. The Solar PV power system will charge a commercial battery bank, to provide energy to the UV water sterilizer on demand 24/7.

This Book includes solar PV power supply examples based on the amount of water you need to sterilize. Operate the UV lamps for four hours each day, or for twenty-four hours for continuous use.

This Book is intended as a Step-by-Step guide to first define your UV Water Sterilizing system, then match that project to one of the Solar Power Supply examples provided. If you need more water treated than the sample systems listed, use Chapter Two to define your project so that your UV Water Sterilizer supplier can quickly identify the right system for your specific project.

Water treatment and sterilization is vital. Water is needed wherever humans operate, and clean potable water can be "produced" onsite from even brackish sources of water. Solar electric (PV) panels are the most effective way to power UV water sterilizers with high performance, reliability, and no fuel-costs at Remote Sites.

Natural disasters, man made emergencies, and remote areas need water treatment wherever humans are based. Solar electric (PV) panels, at historic low prices, lower costs and can be your water UV Sterilizer power-supply solution.

UV water sterilizers use high intensity Ultra-Violet (UV) light to kill the dangerous pathogens that live in natural water supplies. Clean water can be produced from Fresh water, and Salt water sources. This eBook is designed as a solar power guide to sizing, and building your stand-alone, off-grid UV water treatment system, with independent solar power supply.

Clean water is a vital need. Solar PV panels are well suited to provide power for UV water sterilizing systems for remote locations. This Book is written to be a resource in this effort.

Chapter One - UV Water Sterilizers How they Work

Ultraviolet (UV) light has long been known as an ideal method of producing safe drinking water from polluted sources. Many years ago researchers discovered special wavelengths of UV light can kill disease-causing pathogens in our drinking water by attacking the cell DNA rendering the micro-organism inert.

Naturally, or artificially produced, 254 nm UV radiation properly delivered is highly effective at sterilizing water of dangerous pathogens.

UV light, of sufficient dose, as a sterilizer will effectively destroy all common bacteria, viruses and spores which are regularly found in drinking water including Coliform, E. coli, Cryptosporidium,

Hepatitis, Influenza, M. tuberculosis, Giardia, V. cholera , Legionella, Salmonella, B. anthracis, to name a few. UV light as a sterilizer, with proper filtering, kills (99.99%) of the pathogens present in the water, using no chemicals, rendering brackish water clean, potable, and pleasant to drink.

In a Natural, or Man-made disasters, the electrical grid is the first to go. Water and waste treatment, if it existed on site, is often fatally compromised in disasters leaving either no infrastructure, or no power supply available to run it. Off-grid, or stand-alone solar PV power systems can provide power for an individual water treatment system, and have a much greater chance of staying operational in a disaster not being connected to the grid.

UV technology mimics nature to kill disease-causing pathogens in water. Acting just like the UV rays in sunlight, UV rays in UV systems attack the DNA of the pathogens, killing the cells and making your water safe to drink.

UV Water Treatment systems us electrical power to energize a high power UV lamp. This UV lamp is surrounded by a transparent water tube that pushes the water up and around the tube at all angles under the UV irradiation for a given flow-rate.

The energy required by the UV sterilizer system itself is very low as the UV lamp ballasts are very efficient. The low power requirements of UV water sterilizers make them well suited to being solar powered onsite. Solar powered UV water treatment systems are well matched for practical use in remote locations, and as this ebook hopes to show, at great advantage to the installer/operator.

Why Sterilize Water with UV Treatment?

There are many ways to sterilizer water. The dangerous pathogens in water can be destroyed using Ozone, Hydrogen Peroxide, Chlorine, and even Hydroxyl Radicals (OH negative), and, if designed well, are effective. However, none of these approaches has reached the maturity to be as cost-effective in remote sites, and Solar Powered, as UV sterilization has become, in the author's experience.

Ultra-Violet (UV) water treatment and sterilization uses an approach of First filtering out all of the particulates with the sediment filter, or filters. Next, the UV system filters out the remaining particulates (down to 5 Microns) with a Carbon Block filter. Once the particulates are removed, the final stage begins with high-dose UV irradiation. Spiraling up and around the UV Lamp a fine stream of water is irradiated from all angles destroying the micro-organisms to 99.99% removal. UV water treatment systems are self-monitoring, and provide warning alarms if the UV Lamp falls below standard for any reason.

Advantages to UV Treatment for Water Sterilization:

No Chemicals are used in the UV sterilization producing no environmental impacts, no residuals, and no over-dosing possible, as with Chemical treatments. UV technology, using no chemicals, produces no residual chemical byproducts other chemical approaches may introduce, such as the combination of Chlorine and Organics producing trihalomethanes. All of these problems are avoided with UV sterilizing.

UV water sterilizers are best used in "Point of Use" applications. Installed at the "Point of Consumption," the last step in the water treatment process, UV sterilizers offer real-time, and immediate delivery of Drinking water. This "Immediate Treatment" capability insures the water you're delivering is potable and up to standard ready to be used by people.

UV water sterilizers, using 5 Micron Carbon Block filters, has no change in Taste, Odor, pH, or water conductivity. Essential minerals and trace elements remain dissolved in the water producing healthy, taste-free, drinking water on-demand.

UV water sterilizer systems are self-monitoring, and offer Automatic Operation. Easy to install as a factory pre-assembled and tested mountable system, the UV systems listed in the examples below are easy to work with in the field. Replacing filter cartridges, and UV Lamps, when required, is straight forward and easy to do in a few minutes. The UV lamp monitor sounds the alarm if you have any UV lamp issues, so these well designed water sterilizers offer reliability in working conditions.

UV water sterilizer systems are economical to operate. You can expect to sterilizer Hundreds of Gallons per penny of operating costs. Coupled with a solar power supply, your UV water treatment system can be built entirely free from fuel costs. If your site, or location is very remote, not transporting, or buying fuel can be a great advantage.

Chapter Two - Defining Step-by-Step the Best UV Water Treatment System for your Job

Sizing your UV water treatment system is all about delivered Gallons per Day. Reading this book suggests you have a water treatment project in mind. Is your source of water from a well, shallow source, or municipal tap? The following steps will define your Water Sterilizing needs as the basis to choose the best hardware for the job.

Step One: What is the Source of your Water?

The first question becomes "Is your source of water Fresh, or Salty?" Fresh, though brackish, can be wells, puddles, creeks, streams, ponds, lakes, or small rivers. Salty sources of water would be from ocean, or near-ocean sites. If you need to treat Salty water you'll need a Reverse Osmosis (RO) system, needing its own solar power supply, to pre-treat the water before the UV sterilizer.

RO water treatment systems remove salts from the water stream, but they do not guaranty the water is safe to drink. To remove bacteria, and virus, and other pathogens you'll need a UV water sterilizer system. For Salty water sources please visit Chapter 8, as you'll need to include an RO system in your project.

Step Two: What is the Water Pressure of your Water Source?

Your source of water will either have its own pressure, such as with the municipal tap, or elevated water tank, or wont. If your water source is unpressurized you'll need to provide pressure. UV Water Sterilizers require an intake water pressure to work, and have a maximum operating pressure of 125 PSI.

Common water pressure from your municipality varies but usually rates at 30 psi. If your source of water is the Municipal tap then pressure will come from the pressure fed line and you're fine to hook up directly to the UV Water Sterilizer system.

Many remote sites use a Tank, or Cistern, located above the cabin, or house to provide water pressure. This "Gravity" fed system provides pressure for the water line an into the UV water sterilizer. If you're building your tank, be sure to place your tank, or cistern at least 70 feet above the house, in elevation, to provide a satisfactory pressure. This height (70 feet) will provide the nominal 30 PSI pressure you'll need, and enjoy.

If your source of Water is from a Well, you can pump your water and store water in a tank, as described above, or you can connect a separate Solar Water Pump, to pump water from your well directly to your UV water treatment system.

The UV water treatment system will have an In-line filter at the intake to begin to filter the larger particles dissolved in your water such as dirt, rust, and other scale, with your second stage Carbon filter taking out the smaller particulates, and chemicals down to 5 Microns. For more information on Solar Power Supplies for Well pumps, please refer to my eBook "Solar PV Water Pumping."

If your source of water is very shallow, such as a pond, lake, creek, stream, tank, or cistern, you'll need to provide a means of water pressure. One solution is to connect a Surface Pump directly to the UV water sterilizer system.

Connecting a Surface Pump directly to your UV sterilizer allows you to source water for your system from absolutely polluted and brackish sources. Ideal for real world conditions. Surface pumps also have an Inline filter placed before the pump to remove suspended particles.

The UV Sterilizer will also have an Inline filter set for maximum filtering. For more information on Solar Power Supply specifications for Surface Pumps please refer to my eBook "Solar PV Water Pump."

Step Three: What is the Water Quality of my Source Water?"

The source water you are using as feedstock is a key consideration when choosing the right equipment. If your water source is a deep well, then you're in the best situation as deep well water is usually very clean, and may not require additional filtering.

However, if your water source is from a well, you can either store your water in an elevated tank, or you can connect your Submersible well pump directly to the UV Sterilizer. See "Solar PV Water Pumping" for more information on submersible pumps.

If your source of water is from a Shallow surface source, such as a puddle, pond, stream, creek, river, or any kind of surface water, then you'll certainly have particulates, and other pollution present. For surface sources you'll need a Surface Pump to provide pressure to the UV water treatment system. See "Solar PV Water Pumping" for more information on Surface pumps. In all cases Surface sourced water must be filtered.

The UV Sterilizer systems listed here in the examples will have Two-stages of filtration. The first stage is the Sediment phase. In-line filters, are in cartridge form and are rated for particulates down to 5 Microns. The sediment filter takes out the larger particles in the water such as dirt, rust, and other particulates suspended in the water.

The Second stage filter is a Carbon Block filter, and removes chlorine, odors, and tastes, and any other particles which get through Stage-one, also removing particulates down to 5 Microns.

If you are facing a particularly challenging water quality, then add Additional Filters inline. Another Set of Cartridge 10" or 30" Stage one Filters, and Stage two filters will may get your water down to ideal standards.

Turbidity - (Suspended Solids)

The turbidity of your source water is important. Suspended particles in water can encompass, or block UV light from reaching each micro-organism in the water. The Stage-One Sediment filter (5 Micron) will remove dirt, rust, and larger particles. The Stage-Two Carbon Block filter (5 Micron) will sweep up any chlorine and other small bits rendering your water ready for the final UV irradiation stage.

Sample your water and test for turbidity. You'll need to keep your turbidity Less Than 1.0 NTU The Inline filters mentioned above should operate in most conditions to achieve this less than 1.0 NTU rating. If your site water has massive turbidity, then use an additional filter cartridge set In-line as a pre-treatment.

TDS - (Total Dissolved Solids)

Your TDS rating should not exceed 500 ppm. Total Hardness (Calcium and Magnesium) must be Less Than 10 gpg (Grains per Gallon) of hardness. If your sample exceeds this value then a Water Softener device can be added Inline before the filters.

Tannins and Color must Less Than 2 ppm in your sample, or you'll need a pretreatment water softener.

Iron - Must be Less than 0.33 ppm

Manganese - Must be Less than 0.05 ppm

If your sample exceeds any of these standards, you'll need to add filters, or a water softener to act as a Pre-Treatment and pre-clean your incoming water. The On-board Filters (Stage One Sediment, and Stage Two Carbon Block filter) included in your UV treatment system will further filter, then irradiate the water with high-dose UV, rendering the water clean, pleasant, and drinkable.

Step Four: How much water do I need Each Day in Gallons Per Day?"

The size of your Solar Power supply is directly related to how much water you need to sterilize Each Day. The more water you need, the larger your solar PV power system will need to be built.

Residential demands vary with use and lifestyle. Small cottages, cabins, and homes up to 3 people usually need at least 240 Gallons per Day for drinking, cooking, showering etc. This comes to 80 Gallons per person per Day for all uses including showers, cooking, and general consumption, however, you should analyze your actual water needs and come up with a Gallons per Day figure.

Step Five: How Much Solar Energy do I need to Power the UV System?

The total amount of Water each Day you sterilize is the key question in terms of sizing your Solar power supply. The Sample Systems listed below have already been calculated, however if you wish to size your own systems the following information is useful.

UV Water Sterilizers are usually rated in Gallons Per Minute (GPM). As there are 60 minutes per hour each hour of water pumped will be 60 times GPM. If the GPM is 10 Gallons per minute, then one hour would deliver 600 Gallons. Solar electric panels, however, deliver energy over the day, and we estimate how many "Peak" hours equivalent any given location receives from the Sun to calculate how much Energy a given solar PV panel will produce.

The Sun is a powerful source of energy. In terms of peak power in solar energy, the sun is rated at Standard Test Conditions (STC).

The STC condition defines the peak power density of solar energy at the surface of the Earth at 1,000 watts of power per Square Meter (about 10.5 square feet). Note: STC also defines the amount of air-mass the sun path takes (1.5 AMO), standard temperature of 25 degrees C (77 degrees F), the wind speed of 2 meters/sec to further define a standard condition for testing.

To determine how much Solar Energy you have at your location look up the Sun Peak-Hours for your location on a Solar Map. In our examples here we're using a location in Kansas, with 5.5 Solar Peak-hours. Look up your locations solar peak-hour rating.

Raw solar energy produces, at peak condition during a clear sky 1 Kilowatt (1,000) watts of optical power available for conversion. Solar electric modules (Photovoltaic PV Panels) convert this optical energy into Direct Current (DC) with good efficiency delivering about 140 watts of electricity per square meter.

Solar PV panels are "hard wired" to produce a desired voltage. Each solar "Cell" produces about 1/2 Volt DC on its own. Amazingly, even under cloudy conditions solar cells produce good voltages.

The amount of solar energy striking the PV panel will drive the amount of "Current" the solar cells produce. More direct sun, more current produced. Solar cells are interconnected to produce Solar modules which you'll use for your solar UV Water Treatment project.

One square meter of sunlight produces a power electrical force. Producing 140 watts at 12 VDC means over 10 Amps of current is generated. This is a respectable amount of power and can sterilize an amazing amount of water.

Once you know your Water Volume per Day desired for any given UV water sterilizer system project, now you're able to size and power this system with the appropriate solar PV system. In the chapters below we'll go over different UV Water Sterilizer systems for given Flow-rates, and water volumes.

Step Five: Select the Best Solar PV powered Water Treatment System

From the Chapters below, select the best solar PV powered UV system for your project. Match the System Example that most closely matches the Total Amount of Water you wish to deliver Each Day in Gallons Per Day (GPD). Some applications, such as food processing, that may require larger Flow rates. The systems listed below are arranged by Flow-rate, and Total Gallons per Day delivered.

Once you know these vital statistics about your solar UV water treatment project your hardware supplier can know how to configure your system. Your other choice is to match the systems presented in this book that most closely meet your water treatment requirements. If you don't see a system powerful enough listed in this ebook, then go through the steps above, and visit **Solardyne.com** on the worldwide web for more information on larger systems.

Chapter Three: Solar Power Systems using Solar PV Panels Charging Batteries for the Power Supply

The Sun is a powerful source of energy, and ideal for powering remote UV water sterilizer systems. Solar modules produce strong DC currents, and are well suited to extreme locations for their proven durability, and reliability. Solar PV panels produce strong voltages even in low light levels giving you some ability to charge your battery bank even in cloudy weather. Solar PV arrays are configured to provide specified performance over a wide range of climate conditions. Therefore, the solar PV battery charging systems are "oversized" to compensate for variability in location.

UV Water Treatment systems require a power supply. The total "energy" required to power an electrical load is calculated by knowing the power demand, and the hours per day you operate the equipment. Energy, equals Power over Time. One Kilowatt of power, used for one hour, requires one Kilowatt-hour (kWh) of energy.

Natural sunlight contains many wavelengths of light, and can be used, separately, for different purposes. Short wavelengths (from 200-400 nm), like UV are ideal for sterilizing and water treatment uses. Visible wavelengths (from 400-720 nm), from Violet, Indigo, Blue, Green, Yellow, Orange, and Red, getting progressively longer in wavelength, are excellent for Solar Photovoltaic (PV) electricity production.

The Longest wavelengths present in sunlight, the Infra-Red (720-1100 nm), is ideal for thermal applications, such as heating Air, or Water. However, for Water Sterilizing functions, only short UV-B rays (around 254 nm) are capable of destroying micro-organisms in the water.

Direct solar conversion technology exists, and uses the naturally occurring UV part of the spectrum to directly disrupt the dangerous pathogens in the water. Direct use of solar UV is in the experimental (and demonstrable) stage, but not as compact and reliable as running well developed UV sterilizers with solar electricity.

Also, it's interested to note, naturally occurring UV light is less than 2% of the available energy in the solar spectrum. However, our approach here is to use solar energy as an electrical power supply.

Modern solar photovoltaic (PV) panels can be 14% efficient in the field. Therefore, thermodynamically, converting solar energy, first, to electricity and driving a UV lamp, produces many more times 254 nm UV light than occurs in natural sunlight per Square meter.

This Book uses examples of solar energy to produce electricity. Solar electricity is used to charge a battery system. The solar-charged battery will power an inverter to provide standard AC current which, in turn, powers the UV Water Treatment system on demand.

The Solar Power System for your UV sterilizer will include the Solar PV panel array, with the mounting hardware to attach, and deploy your PV panels onsite. The DC electricity from the Solar PV panels is connected to a Charge-Controller.

The Charge-Controller is the "brain" of the system, and performs several functions to keep your power system safe, and operating efficiently. The Charge controller adjusts the power coming from the Solar PV panel by finding it's Maximum Power Point. Controllers use this Maximum Power Point Tracking (MPPT) to match the ideal draw from the PV panels to charge the specific voltage of the batteries.

The Charge-controller, also monitors the battery working voltage, and provides protection for the battery from two conditions: High Voltage, and Low Voltage.

The High Voltage condition is when your batteries are beginning to over-charge. Over-charging is dangerous for batteries, and can lead to failure. Therefore, the charge-controller senses this condition, and employs a High Voltage Disconnect (HVD). The HVD tells the controller to open the circuit from the solar PV panels so that no more charging can occur.

On the other side, if the Battery Voltage is sensed by the controller to be too low, the controller uses a Low Voltage Disconnect (LVD) to turn-off the circuit to power the load, and no more power is drawn from the battery. The LVD condition, is also dangerous to batteries, and is used to further protect the circuit.

Because water treatment is so vital, the user must be able to turn-on the system and have clean water production on demand 24/7. To do this we use a battery bank to store energy from the PV panels and provide power for the UV sterilizer.

The battery bank examples, listed below in the sample systems, are based on the total Energy required by the UV water sterilizer to run for a given number of hours per day, and the total amount of water cleaned and delivered in Gallons per Day.

Regarding Power Supplies, all voltages run "downhill." If you want to power a 12 VDC load from a solar PV panel, you'll need to produce more than 12 VDC in voltage to drive the load either from a solar panel or battery. For a 12 VDC Solar PV panel to produce a higher voltage the manufacturer will wire 36 individual solar cells in series within the module. Wiring the individual solar cells in series "Adds" the voltages producing a nominal 18 VDC.

Under load, which is when you connect the UV sterilizer, the voltage will drop as the solar PV panels drives the system.

Smaller solar PV panels from 60 to 135 watts are usually 12 VDC Panels. If you want larger system voltages wire these panels in series. Two in series for 24 VDC. Four in series for 48 VDC. Larger solar PV panels, from 140 watts - 280 watts are wired at 24 VDC each. Wire two PV panels in series for 48 VDC systems.

The DC Voltage of the Solar PV system is determined by the Inverter you choose to Power the Load. From the Inverter input voltage, you determine your working battery voltage (they should match), and working back from there, you'll know what Voltage to wire your solar array. Again, the Solar DC Voltage will match the Battery Voltage, which in turn, matches the Inverter DC Input Voltage.

Note: When wiring solar PV panels wire in Series to increase Voltage (current remains the same), wire in Parallel to increase Current (voltage remains the same).

The energy produced by your Solar PV panel will be the power rating multiplied by your Daily Solar peak-hour rating for your site.

Check you location with a Solar Power Map, and note how many Solar Peak-Hours of solar radiation your site receives.

Mounting Your Solar PV Panels on location - The Options

Solar panels can be mounted a variety of ways. These options include Pole mounting, Ground mounting, Roof mounting, Passive Tracking, and Active Tracking mounting.

Fixed mounts keep the solar PV panel at a specific Tilt-angle and is adjustable. To increase the output of your Solar PV array you can adjust this angle seasonally to maximize solar exposure. All Solar mounts are mounted to face South when your site is in the Northern Hemisphere, (Note: point your panels North, if you're in the Southern Hemisphere).

PV panels for water pumping need a sturdy and reliable mounting bracket. Solar PV panels can be Pole mounted, either on the Top-of-the-pole, as a masthead, or can be Side-Pole mounted. Side-Pole mounting hardware has a bracket along the bottom and top of the Solar PV panels.

Pole mounting is a great option because it keeps your panels above the ground minimizing ground effects such as increased dust. Also, wiring your panels, once they're mounted on the Mounting Hardware bracket is easier to do as crawling under the solar PV panels (J-Boxes are on the Back of the Panel) is handy.

Pole mounting your solar PV panels also makes installation easier. Smaller Solar PV panels will mount on standard 1.5" Schedule #40 pipe. Site preparation involves auguring a hole, and setting your pole in cement and aggregate.

Larger Solar PV arrays, up to 2,000 watts with Top of Pole mounting, will mount on either 2.5" Schedule #40 pipe, 3.5", or 4.5" pipe for the largest arrays. The examples below will call out the specific diameter of your mounting pipe.

For sturdiness and low cost, you can also Ground Mount your Solar panels. Ground Mounting is an A-Frame rack that allows you to Adjust your Tilt Angle. The general ideal angle for mounting your Solar PV panels is found by taking your Latitude angle of the site, and subtract 15 degrees. Therefore, if your location has a latitude of 45 degrees, the proper tilt angle is 30 degrees as measured from horizontal.

Note: If your site is in a Tropical Location, or a very Cloudy location, the best tilt angle is no angle. Mount your panels flat. This will receive the most "Global" solar radiation, that is both direct, and indirect rays.

You can also mount your solar PV array on your roof, if your roof is near your well site. In most cases this not so, so I'll only mention that option.

Solar energy production is increased if you're always pointing the solar PV panel toward the sun. Tracking hardware does this either in one axis - Morning through Night, or on two-axis (Altitude and Azimuth) which is most accurate.

Trackers are categorized in two types: passive, and active, respectively. Passive tracking such as with the Zomeworks gear has great robustness, and increases Solar PV panel output in energy about 25% on average. Passive-type trackers use uneven heating of internal gasses to self-adjust the panels throughout the day, following the sun. In the morning, these trackers reset to the rising sun and repeat the cycle.

Solar PV power systems work best in direct sunlight. Following the sun's path, solar PV panels increase energy production - power production over time.

Active tracking using Wattsun Active Trackers increases the output of solar PV panels as much as 35%. Using servo motors, and a solar sensor, powered by a separate solar PV array, the Wattsun trackers extract the maximum energy out of your Solar PV array. There is a cost increase for the hardware, but system performance increases dramatically. If your site is very remote, I would recommend no moving parts, and go with Top-Pole mounting requiring no maintenance potential. If you have easy access to your site, or you're in a very small foot-print, active-tracking is a great option for boosting performance.

In the sample systems listed below we'll use two Solar PV panels as examples. For smaller Solar PV panels, rated at 12 VDC each, the Dasol panels of 30, 60, 90, and 135 watt power, respectively are cited. For larger Solar PV panels we'll use the REC line using the popular and widely available 250 Watt module (panel) rated at 24 VDC each.

The Batteries, chosen for the Sample system example Part-Lists below, are Sealed-type, leak-proof, and maintenance free. Sealed Gel Batteries are designed to be rugged, and are reliable. These Batteries can operate in any orientation (upside down not recommended), are manufactured for durability, and ship well. All Solar PV battery charging systems will use the properly sized Charge-Controller, which further protects the Battery Bank for reliable, maintenance-free operation.

The Batteries used in the examples are sealed 12 VDC. For larger systems the batteries are wired in

Series, or Parallel, or Both to match the Input voltage to the Inverter.

An inverter is added to convert the DC capacity of the batteries to AC single-phase electricity to power the UV water treatment system.

Installation and Site Considerations for your Solar PV Power Supply

Your Solar Power System will be likely located some distance from your UV Water Sterilizer system. The UV Water Sterilizer should be mounted indoors if the temperature drops below 4 degrees C. (40 degrees F.) The optimum temperature range for UV sterilizing equipment is between 9 degrees C., and 29 degrees C. The solar PV power system can be mounted up to 200 feet from the location of the UV Water Sterilizer system.

Note: If your Solar PV panels need to sited more than 200 feet away from your Battery Bank, and UV Water Sterilizer system, you can increase the Voltage of your Solar PV array to compensate for the Voltage loss through a longer length of wire. Bring your Solar PV electricity in by wire to your Battery bank where your Charge controller, batteries, and Inverter are located. If your site is in a Very Hot location increasing your Solar Array voltage by adding another panel, or substring of panels, in series to increase the voltage of the PV string.

Remote sites are notorious for logistical difficulties. Often, there is no power, which is the point of this eBook - powering UV water sterilizers with Solar PV power. As such, the sensitive electronics of your solar power system will require protection. Included, in the examples below, are all weather battery boxes, which protect your batteries from the weather, and other environmental externalities. Battery boxes come either insulated, or non-insulated. If you're in a colder climate, then uses insulated. In temperate climates, choose non-insulated. If you're in a hot climate, use insulated.

The Solar PV panels will be Top-of-Pole mounted (other options exist, such as Ground, Roof, and Tracking mounts), to mount the Solar PV array to a Masthead. The masthead hardware fits on top of a vertical steel pipe (from 1.5-4.5" in diameter, Schedule #40 pipe) sunk into the ground for mounting the PV Panels. Larger Solar PV arrays can use Ground Mounts as a stable, and reliable platform as the footings can be secured, important in extreme locations.

The general layout is to mount the UV Water Sterilizer system either at the Water Main input to the structure, or at the Point of Use. The Point of Use, is the most desirable as there is no chance of cross-contamination. If you mount the UV system at your Water Main, then be sure to sterilize the down-stream pipe so the cleaned water can reach the user uncontaminated.

The following Chapters will focus on Specific UV Water Treatment Systems and the corresponding Solar PV Power Supply for a given Daily Water Treatment volume in Gallons Per Day (GPD) delivered.

General Plan:

If your source of Water for water treatment is from a Municipal source, then you'll use the UV Sterilizer System, and the Solar Power Supply.

If your source of Water is from a shallow source, such as from a pond, lake, creek, stream, or same elevation Tank, or Cistern, you'll need a source of pressure, therefore, you'll need a Surface Pump. This eBook covers solar power supplies for the UV water sterilizer system. If you need to solar power your pump see my other eBook "Solar PV Water Pumping" for specifications on solar pump and power supply.

If your source of Water is a Deep Well, then you'll need a Submersible pump, see "Solar PV Water Pumping" for specifications on submersible solar pumps and power supplies.

The following examples discuss the Solar Power Supplies for a given UV Water Treatment Flow rate, and number of Hours per Day the system will operate for a given Water delivery in treated water express in Gallons Per Day.

Chapter Four: UV Water Sterilizer System at 4 GPM with Solar Power Supply from 240 to 5,760 Gallons Per Day

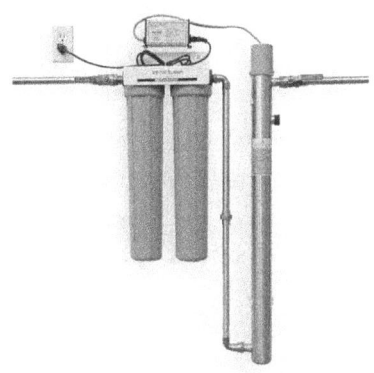

In this chapter we'll look at a UV Water Sterilizing System sized for Small Cottage, or Household use with different Solar Power Supplies based on how much Water per Day you need to sterilize. This UV sterilizing system has a 4 GPM flow rate able to produce 240 Gallons of clean water per Hour. The Total amount of water per Day you can produce depends on the size of the Solar power supply. This UV water treatment system can use tap, ground water, pond, lake, stream, small river, or well sources of water.

The UV Water treatment system used in this example is the Wyckomar SYS-POU250. This UV water treatment system is an All-in-One construction where all of the equipment is pre-assembled, and pre-tested by the manufacturer. The major components in this system include Inline Filters, Filter Housings, UV Lamp Chamber, High-efficiency Ballast with Low Light Alarm, Pressure relief valves, manual Shut-off control, and Intake/Output fittings all mounted on a Stainless steel mounting plate.

The smallest Solar Power Supply in this Chapter will begin with running the UV system for 1 Hour per Day. The next size Solar PV Power Supply will run the system for 2 Hours per Day. The third system size will power the UV sterilizer 4 Hours per Day. The fourth example will run 8 Hours per day, and the last example at continuous duty with a Total Daily output, at the 24 Hour rate, of an estimated 5,760 Gallons Per Day.

Solar Power Supply:

The power demand POS250 UV system is 75 watts. The "energy" demand, therefore, is 75 watt-hours for every hour per day you wish to run the water sterilizer. For this model of UV sterilizer each hour of use will require an additional 75 Watt-Hours of energy, and the system Solar Power supply example becomes larger.

It's easy to construct a Solar PV system to power 12 VDC, or 24 VDC loads, and the examples below will include Parts Lists for each solar power supply. Smaller solar PV systems will be based on a 12 VDC solar battery charging system. The inverter included will convert your battery DC voltage to Single-Phase standard AC current. Your UV Water treatment system is designed for AC electricity, and once both systems, the Solar power and UV sterilizer, are installed: just plug in the UV sterilizer to your Inverter, and turn on.

UV System Pre-Assembled, Pre-Tested and Packaged to Ship

The UV Water treatment system used in the is example is the SYS-POU250 Produced by Wyckomar. This UV system is totally integrated with all of the component sub-systems mounted, tested, and ready to be installed as one unit. Mounted on a stainless steel back panel, this water treatment system is equipped with Two-Stage pre-filters, a UV sterilization lamp Chamber, and monitor with all plumbing, fittings, valves, and system integration.

The SYS-POU250 water sterilizing system is a Point-Of-Use water sterilizer ideal for cottages, RVs, and remote homes, and best installed at the last point in the line before use.

Pressurized Water Source:

If your source of water is from the Municipal tap, pressure tank, or elevated tank, and has water pressure with a minimum of 20 PSI, and a maximum of 125 PSI, then you can connect your UV water sterilizer directly to the water line, either at the Water Main, or at the Point of Use.

Un-pressurized Water Source:

If your source of water is a local Well, then you'll need a Water Pumping System in front of the UV Water Sterilizer to provide the working pressure. If this is the case, please refer to my Book "Solar PV Water Pumping" for specific solar power supplies, and submersible pumps for your particular situation regarding Depth of Well. When selecting your Solar Water Pumping system match your System Flow-Rate to 4 GPM for these examples.

If your water comes from a Shallow Source, such as a pond, lake, creek, stream, or small river, then you'll need a Surface Pump to provide pressure to the UV system. If this is the case, please refer to my eBook "Solar PV Water Pumping" for specific power supplies and pumps for different surface sources of water, including in-line filters which will be necessary. Surface sources of water are typically compromised. Shallow sources of water will require Inline filters (Two-Stage).

Example A - 240 Gallons Per Day

Water sterilizing at 4 GPM - Water delivery rate 240 Gallons Per Hour. Solar Power Supply Run Time: 1 Hour per Day. Total Daily Output in Potable Water Production: 240 Gallons Per Day

Typical Use: Cabins, Boats, RVs, Off-Grid Houses, Remote Sites,

Parts List:

UV Water Sterilizer System:

One (1) SYS-POU250 Wyckomar Water UV Sterilizer System rated at 4 GPM. Includes: 2-Stage water filtrations (5 Micron) Sediment and Carbon filters. High-Intensity UV Lamp, with Quartz Sleeve, and UV Monitor Alarm. Filter Housings, Pressure Relief Valves, with High-efficiency Electronic Ballast. All Pre-assembled, Pre-tested, and mounted to a Stainless steel Mounting Plate

Solar PV Array:

One (1) Solar PV panel rated at 30 watts at 12 VDC. Example solar panel: Dasol DS-A18-30, Size each: 27.2" x 13.8" x 1" One (1) Top-of-Pole Mounting Hardware for one 30 watt panel, or other if vehicle Mounts on 1.5" Schedule #40 pipe.

Battery/Charge-Controller/Inverter:

One (1) SunSaver-6, Charge-controller rated for 12 VDC battery charging up to 6 amps. One (1) Sealed, Maintenance-Free Battery MK 8GU1 rated at 12 VDC @ 31 Amp-hours. One (1) Side-of-Pole Mounted Battery Box (mounted under the Solar PV panels). One (1) ExcelTech XP 125 watt Single-Phase AC Inverter for 12 VDC.

Note: This solar power system is designed to provide One Hour of run time each day for the UV Water sterilizer System producing 240 Gallons per Day of potable water production. Larger water treatment systems listed below.

Example B - 480 Gallons Per Day

Water sterilizing at 4 GPM - Water delivery rate 240 Gallons Per Hour. Solar Power Supply Run Time: 2 Hour per Day. Total Daily Output in Potable Water Production: 480 Gallons Per Day (GPD).

Typical Applications: Cabins, Boats, RVs, Off-Grid Houses, Remote Sites

Parts List:

UV Water Sterilizer System:

One (1) SYS-POU250 Wyckomar Water UV Sterilizer System rated at 4 GPM. Includes: 2-Stage water filtrations (5 Micron) Sediment and Carbon filters. High-Intensity UV Lamp, with Quartz Sleeve, and UV Monitor Alarm. Filter Housings, Pressure Relief Valves, with High-efficiency Electronic Ballast. All Pre-assembled, Pre-tested, and mounted to a Stainless steel Mounting Plate

Solar PV Array:

One (1) Solar PV panel rated at 60 watts at 12 VDC. Example solar panel: Dasol DS-A18-60, Size each: 27.2" x 26.2" x 1.38" One (1) Top-of-Pole Mounting Hardware for one 60 watt panel
Mounts on 1.5" Schedule #40 pipe.

Battery/Charge-Controller/Inverter:

One (1) SunSaver-10, Charge-controller rated for 12 VDC battery charging up to 10 amps. One (1) Sealed, Maintenance-Free Battery MK 8G22NF rated at 12 VDC @ 50 Amp-hours. One (1) Side-of-Pole Mounted Battery Box (mounted under the Solar PV panels). One (1) ExcelTech XP 125 watt Single-Phase AC Inverter for 12 VDC.

Note: This solar power system is designed to provide Two Hours of run time each day for the UV Water sterilizer System. Wire your PV panels in Parallel to increase Amps. System DC voltage: 12 VDC. UV System produces approximately 480 Gallons per Day of potable water production.

Example C - 960 Gallons Per Day

Water sterilizing at 4 GPM - Water delivery rate 240 Gallons Per Hour. Solar Power Supply Run Time: 4 Hour per Day. Total Daily Output in Potable Water Production: 960 Gallons Per Day.

Typical Use: Cabins, Marinas, RVs, Off-Grid Houses, Remote Sites

Parts List:

UV Water Sterilizer System:

One (1) SYS-POU250 Wyckomar Water UV Sterilizer System rated at 4 GPM. Includes: 2-Stage water filtrations (5 Micron) Sediment and Carbon filters. High-Intensity UV Lamp, with Quartz Sleeve, and UV Monitor Alarm. Filter Housings, Pressure Relief Valves, with High-efficiency Electronic Ballast. All Pre-assembled, Pre-tested, and mounted to a Stainless steel Mounting Plate

Solar PV Array:

Two (2) Solar PV panel rated at 60 watts at 12 VDC, 120 Watts total. Example Solar module: Dasol DS-A18-60, Size each: 27.2" x 26.2" x 1.38" One (1) Top-of-Pole Mounting Hardware for two 60 watt panels. Mounts on 1.5" Schedule #40 pipe.

Battery/Charge-Controller/Inverter:

One (1) SunSaver SS-15MPPT, Charge-controller rated for 12 VDC battery charging up to 15 amps. One (1) Sealed, Maintenance-Free Battery MK 8G34 rated at 12 VDC @ 60 Amp-hours each. One (1) Side-of-Pole Mounted Battery Box (mounted under the Solar PV panels). One (1) ExcelTech XP 125 watt Single-Phase AC Inverter for 12 VDC.

Note: DC System. This solar PV system is designed to provide Four Hours of run time each day for the UV Water sterilizer System producing approximately 960 Gallons per Day of potable water production.

Example D - 1,920 Gallons Per Day

Water sterilizing at 4 GPM - Water delivery rate 240 Gallons Per Hour. Solar Power Supply Run Time: 8 Hour per Day. Total Daily Output in Potable Water Production: 1,920 Gallons Per Day.

Typical Use: Cabins, Marinas, Off-Grid Houses, Remote Sites, Restaurants, Wineries, Breweries, Food-processors, Dairy Farms, Cheese factories, Clinics

Parts List:

UV Water Sterilizer System:

One (1) SYS-POU250 Wyckomar Water UV Sterilizer System rated at 4 GPM. Includes: 2-Stage water filtrations (5 Micron) Sediment and Carbon filters. High-Intensity UV Lamp, with Quartz Sleeve, and UV Monitor Alarm. Filter Housings, Pressure Relief Valves, with High-efficiency Electronic Ballast. All Pre-assembled, Pre-tested, and mounted to a Stainless steel Mounting Plate

Solar PV Array:

Two (2) Solar PV panel rated at 135 watts at 12 VDC each, 270 Watt total array. Example PV panel: Dasol DS-A18-135, Size each: 27.2" x 26.2" x 1.38" One (1) Top-of-Pole Mounting Hardware for two 135 watt panels. Mounts on 1.5" Schedule #40 pipe, augured into the ground with cement foundation.

Battery/Charge-Controller/Inverter:

One (1) SunSaver SS-15MPPT, Charge-controller rated for 24 VDC battery charging up to 15 amps. Two (2) Sealed, Maintenance-Free Battery MK 8G34 rated at 12 VDC @ 60 Amp-hours each. One (1) Chest Style Ground Battery Box (can be located up to 50 feet away from PV). One (1) ExcelTech XP/ 24, 125 Watt Single-Phase AC Inverter for 24 VDC.

Note: Two 12 VDC batteries are wired in Series for a 24 VDC system. This solar PV system is designed to provide Eight Hours of run time each day for the UV Water sterilizer System producing approximately 1,920 Gallons per Day of potable water production.

Example E - 5,760 Gallons Per Day

Water sterilizing at 4 GPM - Water delivery rate 240 Gallons Per Hour. Solar Power Supply Run Time: 24 Hours per Day - Continuous Duty. Total Daily Output in Potable Water Production: 5,760 Gallons Per Day.

Typical Use: Cabins, Marinas, Off-Grid Houses, Remote Sites, Residential, Light Commercial, Food-processing, Brewing, Clinics

Parts List:

UV Water Sterilizer System:

One (1) SYS-POU250 Wyckomar Water UV Sterilizer System rated at 4 GPM. Includes: 2-Stage water filtrations (5 Micron) Sediment and Carbon filters. High-Intensity UV Lamp, with Quartz Sleeve, and UV Monitor Alarm. Filter Housings, Pressure Relief Valves, with High-efficiency Electronic Ballast. All Pre-assembled, Pre-tested, and mounted to a Stainless steel Mounting Plate

Solar PV Array:

Four (4) Solar PV panel rated at 250 watts at 24 VDC each, 1,000 Watt total array. Example PV panel: REC Solar PV 250PE, Size each: 65.5" x 39" x 1.5" One (1) Top-of-Pole Mounting Hardware Four (4) 250 watt panels. Mounts on 3.5" Schedule #40 pipe, augured into the ground with cement foundation.

Battery/Charge-Controller/Inverter:

One (1) SunSaver SS-15MPPT, Charge-controller rated for 24 VDC battery charging up to 15 amps. Two (2) Sealed, Maintenance-Free Battery MK 8G30H rated at 12 VDC @ 97 Amp-hours each. One (1) Chest Style Ground Mounted Battery Box (can be located up to 50 feet away from PV). One (1) ExcelTech XP/24, 125 watt Single-Phase AC Inverter for 24 VDC.

Note: Two 12 VDC batteries are wired in series for a 24 VDC system. This solar PV system is designed to provide 24 Hours of run time each day for the UV Water sterilizer System producing approximately 5,760 Gallons per Day of potable water production.

Chapter Five - UV Water Treatment at 8 GPM with Solar Power Supply for 960 to 11,520 Gallons Per Day

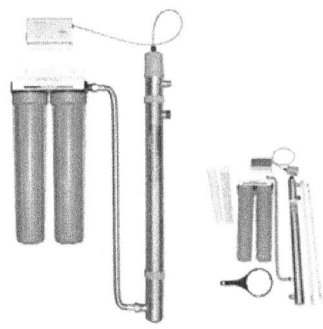

In this chapter we'll look at solar PV powered water treatment system rated at 8 Gallons Per Minute flow rate. Ideal for Residential systems, the UV water treatment system used these examples is the Wyckomar SYS-MD1003. This UV treatment system is build All-in-one which includes all of the necessary hardware pre-assembled and tested. UV treatment systems include inline Two-Stage filtering (5 Micron), Housings, UV Lamp chamber, Quartz Sleeve, Fittings, and Pressure Relief Valves, all installed and ready to go.

The following Solar PV Power supplies are designed to run the MD1003 UV Treatment system for the number of hours specified to deliver a given amount of Drinkable, pleasant, potable water per Day.

Solar Power Supply

The power demand of this system is 95 watts. The "energy" demand, therefore, is 95 watt-hours for every hour per day you wish to run the water sterilizer. For this model of UV sterilizer each hour of use will require an additional 95 Watt-Hours of energy from the Solar PV power system, and the system example get larger.

Example F - 960 Gallons Per Day (GPD)

Water sterilizing at 8 GPM - Water delivery rate 480 Gallons Per Hour. Solar Power Supply Run Time: 2 Hours per Day. Total Daily Output in Potable Water Production: 960 Gallons Per Day.

Typical Use: Cabins, Marinas, Off-Grid Houses, Remote Sites, Residential, Commercial, Food-processing, Brewing,

Parts List:

UV Water Sterilizer System:

One (1) SYS-MD1003 Wyckomar Water UV Sterilizer System rated at 8 GPM. Includes: 2-Stage water filtrations (5 Micron) Sediment and Carbon filters. High-Intensity UV Lamp, with Quartz Sleeve, and UV Monitor Alarm. Filter Housings, Pressure Relief Valves, with High-efficiency Electronic Ballast. All Pre-assembled, Pre-tested, and mounted to a Stainless steel Mounting Plate

Solar PV Array:

One (1) Solar PV panel rated at 135 watts at 12 VDC each. Example solar PV panel: Dasol DS-A18-135, Size each: 27.2" x 26.2" x 1.38" One (1) Top-of-Pole Mounting Hardware for one 60 watt panel. Mounts on 1.5" Schedule #40 pipe.

Battery/Charge-Controller/Inverter:

One (1) SunSaver SS-15MPPT, Charge-controller rated for 12 VDC battery charging up to 15 amps. One (1) Sealed, Maintenance-Free Battery MK 8G24DT rated at 12 VDC @ 73 Amp-hours. One (1) Side-of-Pole Mounted Battery Box (mounted under the Solar PV panels). One (1) ExcelTech XP 125 watt Single-Phase AC Inverter for 12 VDC.

Note: This solar power system is designed to provide Two Hours of run time each day for the UV Water sterilizer System. Wire your PV panels in Parallel to increase Amps. System DC voltage: 12 VDC. UV System produces approximately 960 Gallons per Day of potable water production.

Example G - 1,920 Gallons Per Day (GPD)

Water sterilizing at 8 GPM - Water delivery rate 480 Gallons Per Hour. Solar Power Supply Run Time: 4 Hours per Day. Total Daily Output in Potable Water Production: 1,920 Gallons Per Day.

Typical Use: Cabins, Marinas, Off-Grid Houses, Remote Sites, Residential, Commercial, Food-processing, Brewing, Clinics,

Parts List:

UV Water Sterilizer System:

One (1) SYS-MD1003 Wyckomar Water UV Sterilizer System rated at 8 GPM. **Includes**: 2-Stage water filtrations (5 Micron) Sediment and Carbon filters. High-Intensity UV Lamp, with Quartz Sleeve, and UV Monitor Alarm. Filter Housings, Pressure Relief Valves, with High-efficiency Electronic Ballast. All Pre-assembled, Pre-tested, and mounted to a Stainless steel Mounting Plate.

Solar PV Array:

Two (2) Solar PV panel rated at 135 watts at 12 VDC each, 270 Watt total array. Example PV Panel: Dasol DS-A18-135, Size each: 27.2" x 26.2" x 1.38" One (1) Top-of-Pole Mounting Hardware for two 60 watt panels. Mounts on 1.5" Schedule #40 pipe.

Battery/Charge-Controller/Inverter:

One (1) SunSaver SS-15MPPT, Charge-controller rated for 24 VDC battery charging up to 15 Amps. Two (2) Sealed, Maintenance-Free Batteries MK 8G34 rated at 12 VDC @ 60 Amp-hours each. One (1) Side-of-Pole Mounted Battery Box (mounted under the Solar PV panels). One (1) ExcelTech XP 125 watt Single-Phase AC Inverter for 24 VDC.

Note: DC System wire solar PV panels in Parallel. This solar PV system is designed to provide Four Hours of run time each day for the UV Water sterilizer System producing approximately 1,920 Gallons per Day of potable water production.

Example H - 3,840 Gallons Per Day (GPD)

Water sterilizing at 8 GPM - Water delivery rate 480 Gallons Per Hour . Solar Power Supply Run Time: 8 Hours per Day. Total Daily Output in Potable Water Production: 3,840 Gallons Per Day.

Typical Use: Cabins, Marinas, Off-Grid Houses, Remote Sites, Residential, Commercial, Food-processing, Brewing, Clinics

Parts List:

UV Water Sterilizer System:

One (1) SYS-MD1003 Wyckomar Water UV Sterilizer System rated at 8 GPM. Includes: 2-Stage water filtrations (5 Micron) Sediment and Carbon filters. High-Intensity UV Lamp, with Quartz Sleeve, and UV Monitor Alarm. Filter Housings, Pressure Relief Valves, with High-efficiency Electronic Ballast. All Pre-assembled, Pre-tested, and mounted to a Stainless steel Mounting Plate.

Solar PV Array:

Two (2) Solar PV panel rated at 250 watts at 24 VDC each, 500 Watt total array. Example: REC Solar PV 250PE, Size each: 65.5" x 39" x 1.5" One (1) Top-of-Pole Mounting Hardware for two 250 watt panels. Mounts on 2.5" Schedule #40 pipe, augured into the ground with cement foundation

Battery/Charge-Controller/Inverter:

One (1) MorningStar TS-MTTP-45, Charge-controller rated for 24 VDC battery charging. Two (2) Sealed, Maintenance-Free Battery MK 8G24DT rated at 12 VDC @ 73 Amp-hours each. One (1) Chest Style Ground Mounted Battery Box (can be located up to 50 feet away from PV). One (1) ExcelTech XP/24 125 watt Single-Phase AC Inverter for 24 VDC.

Note: Two 12 VDC batteries are wired in series for a 24 VDC system. Two PV panels wired in Parallel. This solar PV system is designed to provide Eight

Hours of run time each day for the UV Water sterilizer System producing approximately 3,840 Gallons per Day of potable water production.

Example I - 11,520 Gallons Per Day (GPD)

Water sterilizing at 8 GPM - Water delivery rate 480 Gallons Per Hour
Solar Power Supply Run Time: 24 Hours per Day - Continuous Duty
Total Daily Output in Potable Water Production: 11,520 Gallons Per Day

Typical Use: Cabins, Marinas, Off-Grid Houses, Remote Sites, Residential, Commercial, Food-processing, Brewing, Clinics, Hospitals, Small Villages, Farms, Ranches

Parts List:

UV Water Sterilizer System:

One (1) SYS-MD1003 Wyckomar Water UV Sterilizer System rated at 8 GPM. Includes: 2-Stage water filtrations (5 Micron) Sediment and Carbon filters. High-Intensity UV Lamp, with Quartz Sleeve, and UV Monitor Alarm. Filter Housings, Pressure Relief Valves, with High-efficiency Electronic Ballast. All Pre-assembled, Pre-tested, and mounted to a Stainless steel Mounting Plate.

Solar PV Array:

Six (6) Solar PV panels rated at 250 watts at 24 VDC each, 1,500 Watt total array. Example: REC Solar PV 250PE, Size each: 65.5" x 39" x 1.5" One (1) Top-of-Pole Mounting Hardware for six (6) 250 watt panels. Mounts on 3.5" Schedule #40 pipe, augured into the ground with cement foundation.

Battery/Charge-Controller/Inverter:

One (1) MorningStar TS-MPPT-60, Charge-controller rated for 24 VDC battery charging. Two (2) Sealed, Maintenance-Free Battery MK 8G30H rated at 12 VDC @ 97 Amp-hours each. One (1) Chest Style Ground Mounted Battery Box (can be located up to 50 feet away from PV). One (1) ExcelTech XP/24 125 watt Single-Phase AC Inverter for 24 VDC.

Note: Two 12 VDC batteries are wired in series for a 24 VDC system. Solar PV panels wired as two substrings in series. Each substring of 3 panels wired in parallel This solar PV system is designed to provide 24 Hours of run time each day for the UV Water sterilizer System producing approximately 11,520 Gallons per Day of potable water production.

Chapter Six - UV Water Treatment System at 12 GPM for 2,880 GPD to 17,280 Gallons Per Day

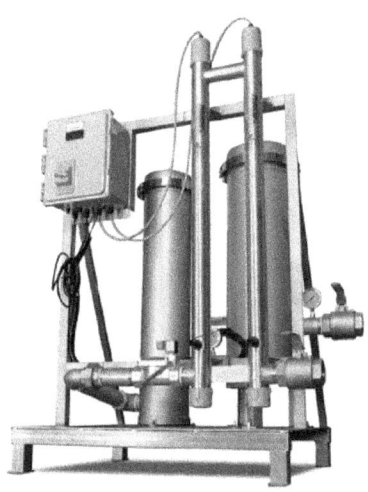

This chapter looks at a higher flow rate UV water sterilizer system the SYS-MD1004. Rated at 12 GPM this UV sterilizer is designed for households, buildings with 1" water lines. The 1" intake provides the increased capacity and can be run for short times each day, or 24 hours per day in continuous use. The Solar Power systems listed below use Solar PV panels to build a Solar Array of the proper power rating. Systems include the mounting hardware suggested, as well as the Charge controller, Battery

Bank, and Inverter to produce the AC power to run your UV sterilizer system.

THe UV Dose from this UV sterilizer produces 54 mJ/cm2 (54,000 µsec/cm2) @ 95% UVT 38 mJ/cm2 (38,000 µsec/cm2) @ 70% UVT. This high dose UV irradiation sterilizes commercial outputs useful for food processing, cheese factories, hospitals, small villages, any installed capacity up to 17,280 GPD in continuous operation.

Example J - 2,880 Gallons Per Day (GPD)

Water sterilizing at 12 GPM - Water delivery rate 720 Gallons Per Hour. Solar Power Supply Run Time: 4 Hours per Day. Total Daily Output in Potable Water Production: 2,880 Gallons Per Day.

Typical Use: Cabins, Marinas, Off-Grid Houses, Remote Sites, Residential, Commercial, Food-processing, Brewing, Clinics, other remote sites.

Parts List:

UV Water Sterilizer System:

One (1) SYS-MD1004 Wyckomar Water UV Sterilizer System rated at 12 GPM. Includes: 2-Stage water filtrations (5 Micron) Sediment and Carbon filters. High-Intensity UV Lamp, with Quartz Sleeve, and UV

Monitor Alarm. Filter Housings, Pressure Relief Valves, with High-efficiency Electronic Ballast. All Pre-assembled, Pre-tested, and mounted to a Stainless steel Mounting Plate for installation ease.

Solar PV Array:

One (1) Solar PV panel rated at 250 watts at 24 VDC. Example: REC Solar PV Panel 250PE, Size: 65.5" x 39" x 1.5" One (1) Top-of-Pole Mounting Hardware for two 250 watt panels. Mounts on 2.5" Schedule #40 pipe, augured into the ground with cement foundation.

Battery/Charge-Controller/Inverter:

One (1) SunSaver SS-15MPPT, Charge-controller rated for 24 VDC battery charging up to 15 amps. Two (2) Sealed, Maintenance-Free Battery MK 8G24DT rated at 12 VDC @ 73 Amp-hours each. One (1) Chest Style Ground Mounted Battery Box (can be located up to 50 feet away from PV). One (1) ExcelTech XP/24 125 watt Single-Phase AC Inverter for 24 VDC.

Note: Two 12 VDC batteries are wired in series for a 24 VDC system. This solar PV system is designed to provide Four Hours of run time each day for the UV Water sterilizer System producing approximately 2,880 Gallons per Day of potable water production.

Example K - 5,760 Gallons Per Day (GPD)

Water sterilizing at 12 GPM - Water delivery rate 720 Gallons Per Hour. Solar Power Supply Run Time: 8 Hours per Day. Total Daily Output in Potable Water Production: 5,760 Gallons Per Day.

Typical Use: Cabins, Marinas, Off-Grid Houses, Remote Sites, Residential, Commercial, Food-processing, Brewing, Clinics, farms

Parts List:

UV Water Sterilizer System:

One (1) SYS-MD1004 Wyckomar Water UV Sterilizer System rated at 12 GPM. Includes: 2-Stage water filtrations (5 Micron) Sediment and Carbon filters. High-Intensity UV Lamp, with Quartz Sleeve, and UV Monitor Alarm. Filter Housings, Pressure Relief Valves, with High-efficiency Electronic Ballast. All Pre-assembled, Pre-tested, and mounted to a Stainless steel Mounting Plate.

Solar PV Array:

Two (2) Solar PV panel rated at 250 watts at 24 VDC each, 500 Watt total array. Example: REC Solar PV 250PE, Size each: 65.5" x 39" x 1.5" One (1) Top-of-Pole Mounting Hardware for two 250 watt panels.

Mounts on 3.5" Schedule #40 pipe, augured into the ground with cement foundation.

Battery/Charge-Controller/Inverter:

One (1) MorningStart TX-MPPT-45, Charge-controller rated for 24 VDC battery charging. Two (2) Sealed, Maintenance-Free Battery MK 8G24DT rated at 12 VDC @ 73 Amp-hours each. One (1) Chest Style Ground Mounted Battery Box (can be located up to 50 feet away from PV). One (1) ExcelTech XP/24 125 watt Single-Phase AC Inverter for 24 VDC input voltage.

Note: Two 12 VDC batteries are wired in series for a 24 VDC system. Solar PV panels wired in Parallel. This solar PV system is designed to provide Eight Hours of run time each day for the UV Water sterilizer System producing approximately 5,760 Gallons per Day of potable water production.

Example L - 8,640 Gallons Per Day (GPD)

Water sterilizing at 12 GPM - Water delivery rate 720 Gallons Per Hour. Solar Power Supply Run Time: 12 Hours per Day. Total Daily Output in Potable Water Production: 8,640 Gallons Per Day.

Typical Use: Cabins, Marinas, Off-Grid Houses, Remote Sites, Residential, Commercial, Food-processing, Brewing, Clinics,

Parts List:

UV Water Sterilizer System:

One (1) SYS-MD1004 Wyckomar Water UV Sterilizer System rated at 12 GPM. Includes: 2-Stage water filtrations (5 Micron) Sediment and Carbon filters. High-Intensity UV Lamp, with Quartz Sleeve, and UV Monitor Alarm, Filter Housings, Pressure Relief Valves, with High-efficiency Electronic Ballast. All Pre-assembled, Pre-tested, and mounted to a Stainless steel Mounting Plate.

Solar PV Array:

Four (4) Solar PV panel rated at 250 watts at 24 VDC each, 1,000 Watt total array. Example: REC Solar PV 250PE, Size each: 65.5" x 39" x 1.5" One (1) Top-of-Pole Mounting Hardware for four 250 watt panels. Mounts on 3.5" Schedule #40 pipe, augured into the ground with cement foundation.

Battery/Charge-Controller/Inverter:

One (1) MorningStar TS-MPPT-45, Charge-controller rated for 24 VDC battery charging. Two (2) Sealed, Maintenance-Free Battery MK 8G27 rated at 12 VDC @ 86 Amp-hours each. One (1) Chest Style Ground Mounted Battery Box (can be located up to 50 feet away from PV). One (1) ExcelTech XP/24 125 watt Single-Phase AC Inverter for 24 VDC input DC voltage.

Note: Two 12 VDC batteries are wired in series for a 24 VDC system. This solar PV system is designed to provide 12 Hours of run time each day for the UV Water sterilizer System producing approximately 8,640 Gallons per Day of potable water production.

Example M - 17,280 Gallons Per Day (GPD)

Water sterilizing at 12 GPM - Water delivery rate 720 Gallons Per Hour. Solar Power Supply Run Time: 24 Hours per Day. Total Daily Output in Potable Water Production: 17,280 Gallons Per Day.

Typical Use: Cabins, Marinas, Off-Grid Houses, Remote Sites, Residential, Commercial, Food-processing, Brewing, Clinics, Hospitals

Parts List:

UV Water Sterilizer System:

One (1) SYS-MD-1004 Wyckomar Water UV Sterilizer System rated at 12 GPM. Includes: 2-Stage water filtrations (5 Micron) Sediment and Carbon filters, High-Intensity UV Lamp, with Quartz Sleeve, and UV Monitor Alarm, Filter Housings, Pressure Relief Valves, with High-efficiency Electronic Ballast. All Pre-assembled, Pre-tested, and mounted to a Stainless steel Mounting Plate

Solar PV Array:

Eight (8) Solar PV panel rated at 250 watts at 24 VDC each, 2,000 Watt total array. Example: REC Solar PV 250PE, Size each: 65.5" x 39" x 1.5" One (1) Top-of-Pole Mounting Hardware for Eight (8) 250 watt panels. Mounts on 6" Schedule #40 pipe, augured into the ground with cement foundation.

Battery/Charge-Controller/Inverter:

One (1) MorningStar TS-MPPT-60, Charge-controller rated for 24 VDC battery charging. Four (4) Sealed, Maintenance-Free Battery MK 8G27 rated at 12 VDC @ 86 Amp-hours each. One (1) Chest Style Ground Mounted Battery Box (can be located up to 50 feet away from PV). One (1) ExcelTech XP/24 125 watt Single-Phase AC Inverter for 24 VDC input.

Note: Four (4) 12 VDC batteries are wired 2 in Parallel, and those substrings wired in series for a 24 VDC system. This solar PV system is designed to provide 24 Hours of run time each day for the UV Water sterilizer System producing approximately 17,280 Gallons per Day of potable water production.

Chapter Seven - UV Water Sterilizing Systems for 30 GPM for 7,200 to 43,200 Gallons Per Day

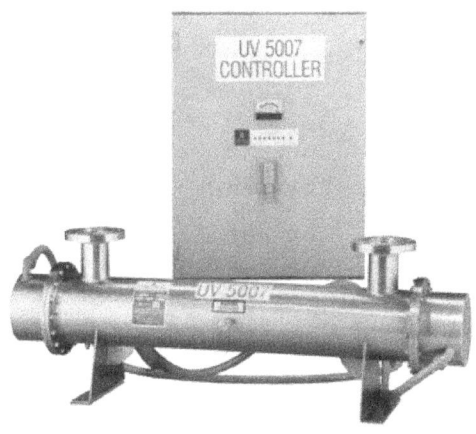

Large UV Water Treatment systems have a large appetite for source water and power. The SYS-MD-1006 rated at 30 GPM. Sized for 1.5" intake pipes this commercial unit can process up to 43,200 Gallons per Day. The MD-1006 is a large commercial UV water treatment system. The Water Intake pipe is 1.5" in diameter.

Example N - 7,200 Gallons Per Day

Water sterilizing at 30 GPM - Water delivery rate 1,800 Gallons Per Hour. Solar Power Supply Run Time: 4 Hours per Day. Total Daily Output in Potable Water Production: 7,200 Gallons Per Day.

Typical Use: Cabins, Marinas, Off-Grid Houses, Remote Sites, Residential, Commercial, Food-processing, Brewing, Clinics, Hospitals, Small Villages

Parts List:

UV Water Sterilizer System:

One (1) SYS-MD-1006 Wyckomar Water UV Sterilizer System rated at 30 GPM. Includes: 2-Stage water filtrations (5 Micron) Sediment and Carbon filters, High-Intensity UV Lamp, with Quartz Sleeve, and UV Monitor Alarm, Filter Housings, Pressure Relief Valves, with High-efficiency Electronic Ballast. All Pre-assembled, Pre-tested, and mounted to a Stainless steel Mounting Plate.

Solar PV Array:

Two (2) Solar PV panel rated at 250 watts at 24 VDC each, 500 Watt total array. Example: REC Solar PV 250PE, Size each: 65.5" x 39" x 1.5" One (1) Top-of-Pole Mounting Hardware for two 250 watt panels. Mounts on 2.5" Schedule #40 pipe, augured into the ground with cement foundation.

Battery/Charge-Controller/Inverter:

One (1) MorningStar TS-MPPT-45, Charge-controller rated for 24 VDC battery charging. Two (2) Sealed, Maintenance-Free Battery MK 8G34 rated at 12 VDC @ 60 Amp-hours each. One (1) Chest Style Ground Mounted Battery Box (can be located up to 50 feet away from PV). One (1) ExcelTech XP/24 125 watt Single-Phase AC Inverter for 24 VDC.

Note: Two 12 VDC batteries are wired in series for a 24 VDC system. This solar PV system is designed to provide Four Hours of run time each day for the UV Water sterilizer System producing approximately 7,200 Gallons per Day of potable water production.

Example O - 14,400 Gallons Per Day

Water sterilizing at 30 GPM - Water delivery rate 1,800 Gallons Per Hour. Solar Power Supply Run Time: 8 Hours per Day. Total Daily Output in Potable Water Production: 3,600 Gallons Per Day.

Typical Use: Cabins, Marinas, Off-Grid Houses, Remote Sites, Residential, Commercial, Food-processing, Brewing, Clinics, Hospitals, Food-processors, Wineries, Breweries, Restaurants.

Parts List:

UV Water Sterilizer System:

One (1) SYS-MD-1006 Wyckomar Water UV Sterilizer System rated at 30 GPM. Includes: 2-Stage water filtrations (5 Micron) Sediment and Carbon filters. High-Intensity UV Lamp, with Quartz Sleeve, and UV Monitor Alarm. Filter Housings, Pressure Relief Valves, with High-efficiency Electronic Ballast. All Pre-assembled, Pre-tested, and mounted to a Stainless steel Mounting Plate.

Solar PV Array:

Four (4) Solar PV panel rated at 250 watts at 24 VDC each, 1,000 Watt total array. Example PV panel: REC Solar PV 250PE, Size each: 65.5" x 39" x 1.5" One (1) Top-of-Pole Mounting Hardware for four (4) 250 watt panels. Mounts on 3.5" Schedule #40 pipe, augured into the ground with cement foundation.

Battery/Charge-Controller/Inverter:

One (1) MorningStar TS-MPPT-60, Charge-controller rated for 24 VDC battery charging up to 10 amps. Two (2) Sealed, Maintenance-Free Battery MK 8G30H rated at 12 VDC @ 97 Amp-hours each. One (1) Chest Style Ground Mounted Battery Box (can be located up to 50 feet away from PV). One (1) ExcelTech XP/24 125 watt Single-Phase AC Inverter for 24 VDC input voltage.

Note: Two 12 VDC batteries are wired in series for a 24 VDC system. This solar PV system is designed to provide Eight Hours of run time each day for the UV Water sterilizer System producing approximately 17,280 Gallons per Day of potable water production.

Example P - 21,600 Gallons Per Day

Water sterilizing at 30 GPM - Water delivery rate 1,800 Gallons Per Hour. Solar Power Supply Run Time: 12 Hours per Day. Total Daily Output in Potable Water Production: 21,600 Gallons Per Day.

Typical Use: Cabins, Marinas, Off-Grid Houses, Remote Sites, Residential, Commercial, Food-processing, Brewing, Clinics, Hospitals, Small Villages.

Parts List:

UV Water Sterilizer System:

One (1) SYS-MD-1006 Wyckomar Water UV Sterilizer System rated at 30 GPM. Includes: 2-Stage water filtrations (5 Micron) Sediment and Carbon filters, High-Intensity UV Lamp, with Quartz Sleeve, and UV Monitor Alarm, Filter Housings, Pressure Relief Valves, with High-efficiency Electronic Ballast. All Pre-assembled, Pre-tested, and mounted to a

Stainless steel Mounting Plate for one piece installation.

Solar PV Array:

Six (6) Solar PV panel rated at 250 watts at 24 VDC each, 1, 500 Watt total array. Example PV panel: REC Solar PV 250PE, Size each: 65.5" x 39" x 1.5" One (1) Top-of-Pole Mounting Hardware for six (6) 250 watt panels. Mounts on 6" Schedule #40 pipe, augured into the ground with cement foundation

Battery/Charge-Controller/Inverter:

One (1) MorningStar XS-MPPT-45, Charge-controller rated for 24 VDC battery charging. Two (2) Sealed, Maintenance-Free Battery MK 8G30H rated at 12 VDC @ 97 Amp-hours each. One (1) Chest Style Ground Mounted Battery Box (can be located up to 50 feet away from PV). One (1) ExcelTech XP/24 125 watt Single-Phase AC Inverter for 24 VDC.

Note: Two 12 VDC batteries are wired in series for a 24 VDC system. This solar PV system is designed to provide 12 Hours of run time each day for the UV Water sterilizer System producing approximately 21,600 Gallons per Day of potable water production.

Example Q - 43,200 Gallons Per Day

Water sterilizing at 30 GPM - Water delivery rate 1,800 Gallons Per Hour. Solar Power Supply Run Time: 24 Hours per Day - Continuous Duty. Total Daily Output in Potable Water Production: 43,200 Gallons Per Day.

Typical Use: Cabins, Marinas, Off-Grid Houses, Remote Sites, Residential, Commercial, Food-processing, Brewing, Clinics, Small Villages

Parts List:

UV Water Sterilizer System:

One (1) SYS-MD-1006 Wyckomar Water UV Sterilizer System rated at 30 GPM. Includes: 2-Stage water filtrations (5 Micron) Sediment and Carbon filters. High-Intensity UV Lamp, with Quartz Sleeve, and UV Monitor Alarm. Filter Housings, Pressure Relief Valves, with High-efficiency Electronic Ballast. All Pre-assembled, Pre-tested, and mounted to a Stainless steel Mounting Plate for easy installation.

Solar PV Array:

Eight (8) Solar PV panel rated at 250 watts at 24 VDC each, 2,000 Watt total array. Example PV panel: REC Solar PV 250PE, Size each: 65.5" x 39" x 1.5" One (1) Top-of-Pole Mounting hardware for eight (8) 250 watt panels Mounts on 6" Schedule #40 pipe, augured into the ground with cement foundation.

Battery/Charge-Controller/Inverter:

One (1) MorningStar TS-MPPT-60, Charge-controller rated for 24 VDC battery charging. Four (4) Sealed, Maintenance-Free Battery MK 8G30H rated at 12 VDC @ 97 Amp-hours each
One (1) Chest Style Ground Mounted Battery Box (can be located up to 50 feet away from PV)
One (1) ExcelTech XP/24 125 watt Single-Phase AC Inverter for 24 VDC

Note: Four 12 VDC batteries are wired as 2 battery substrings in Parallel, those substrings in series for a 24 VDC system. This solar PV system is designed to provide 24 Hours of run time each day for the UV Water sterilizer System producing approximately 43,200 Gallons per Day of potable water production.

Chapter Eight: Quick Guide to UV Water Treatment System Examples by Flow-Rate, and Gallons per Day

In each chapter above, are listed different Solar PV powered UV Water Treatment systems, based on whether you're pumping from a Well, or from a Shallow source. Examples are defined by Flow Rates, and Daily water delivery in Gallons Per Day. Browse the systems below and match your project specifications, and needs, to the system listed that comes closest to your water requirements.

Examples of Solar PV powered UV Water Treatment Systems by Flow rate in Gallons per Minute (GPM), and Total Daily Gallons in Gallons per Day (GPD):

System A: 4 GPM, Delivers 240 GPD

System B: 4 GPD, Delivers 480 GPD

System C: 4 GPD, Delivers 960 GPD

System D: 4 GPD, Delivers 1,920 GPD

System E: 4 GPD, Delivers 5,760 GPD

System F: 8 GPD, Delivers 960 GPD

System G: 8 GPD, Delivers 1,920 GPD

System H: 8 GPD, Delivers 3,840 GPD

System I: 8 GPD, Delivers 11,520 GPD

System J: 8 GPD, Delivers 2,880 GPD

System K: 8 GPD, Delivers 5,760 GPD

System L: 12 GPD, Delivers 8,640 GPD

System M: 12 GPD, Delivers 17,280 GPD

System N: 30 GPD, Delivers 7,200 GPD

System O: 30 GPD, Delivers 14,400 GPD

System P: 30 GPD, Delivers 21,600 GPD

System Q: 30 GPD, Delivers 43,200 GPD

Be sure to plan your solar PV powered UV water treatment project in terms of Site-Preparation, UV Water Treatment Equipment Installation, Solar Power Supply, and all cables, piping, and grounding.

Always use **CAUTION** when installing electrical devices. Solar PV panels produce respectable voltages and currents and all safety procedures should be followed. Be sure to Read your Installation Manual carefully, and follow the instructions to the letter.

Properly installed, and maintained, solar PV powered UV Water Treatment systems offer long life, great productivity, and ease of installation and operation. For more information on UV water treatment, solar PV panels, Batteries, Inverters, Charge controllers, or other hardware please visit **Solardyne.com** on the Worldwide Web.

Enjoy your Solar Water Treatment project!